Alan McKirdy has written many popular books and book chapters on geology and related topics and has helped to promote the study of environmental geology in Scotland. His other books with Birlinn include *Set in Stone: The Geology and Landscapes of Scotland* and *Land of Mountain and Flood*, which was nominated for the Saltire Research Book of the Year award. Before his retirement, he was Head of Knowledge and Information Management at Scottish Natural Heritage. Alan is now a freelance writer and has given many talks on Scottish geology and landscapes at book festivals and other events across the country.

Edinburgh

LANDSCAPES IN STONE

Alan McKirdy

For Stuart Monro

First published in Great Britain in 2017 by
Birlinn Ltd
West Newington House
10 Newington Road
Edinburgh
EH9 1QS

www.birlinn.co.uk

ISBN: 978 1 78027 371 6

Copyright © Alan McKirdy 2017

The right of Alan McKirdy to be identified as the author of this work has been asserted by him in accordance with the Copyright, Designs and Patents Act, 1988

All rights reserved. No part of this publication may be reproduced, stored, or transmitted in any form, or by any means, electronic, mechanical or photocopying, recording or otherwise, without the express written permission of the publisher.

British Library Cataloguing-in-Publication Data
A catalogue record for this book is available on request from the British Library

Designed and typeset by Mark Blackadder

FRONTISPIECE.
Aerial view of Castle Rock and Edinburgh city.

Printed and bound in Britain by Latimer Trend, Plymouth

Contents

Introduction — 7

Edinburgh through time — 8

Geological map — 10

1. Edinburgh – birthplace of modern geology — 11
2. It's a dynamic Earth — 15
3. A disappeared ocean and baking deserts — 16
4. Roaring volcanoes and tropical lagoons — 18
5. A cauldron of steam and lava — 26
6. Tropical seas and swamps — 29
7. Mind the gap! — 32
8. The Ice Age — 33
9. Building the New Town — 38
10. Places to visit — 41

Acknowledgements and picture credits — 48

Introduction

An ancient and long-extinct volcano lies at the centre of Scotland's capital. It roared into life some 350 million years ago and has been a source of fascination since it was first studied in earnest during the Scottish Enlightenment – Scotland's time as cultural and scientific doyen of the western world. Dr James Hutton, the acknowledged founder of modern geology, was at the heart of this new way of looking at the world. Many of his ground-breaking ideas of how the world works were predicated on the rocks and landscapes of his home city and surrounding area. It can be said without fear of contradiction that, as a result of his seminal work and those who came after him, Scotland gave the study of geology to the world.

The architecture of Edinburgh is also a source of huge civic pride, and rocks quarried locally from ancient, now long disappeared, rivers and seas provided the architects and builders of the early nineteenth century with the required raw materials to create the stunning elegance of Edinburgh's New Town. Coal deposits and oil shale were also exploited from the Industrial Revolution to recent times, powering many aspects of Edinburgh's development.

During the last two million years, the landscape was shaped and moulded by the passage of ice during a great freeze that left an indelible stamp on Edinburgh's cityscape. The position of the coastline that defines the northern edge of the Lothians has changed in rhythm with the prevailing climate. But further change will affect many aspects of the cityscapes and wider environment of Edinburgh as the consequences of climate change really begin to bite.

Opposite. Salisbury Crags.

Edinburgh through time

Period of geological time	Millions of years ago	Scotland's global position	Environments and events in and around Edinburgh
Anthropocene	Last 10,000 years	57° N	After the ice melted, the land rose relative to the sea, so many coastal areas are now wide rock platforms abandoned by the sea. These 'raised beaches' are impressive and obvious landscape features today. But this time is primarily the age of *Homo sapiens* – us!
Quaternary	Started 2 million years ago	Present position of 57° N	This was the age of ice. Many advances and retreats of the ice are recorded during this period. Glaciers created the familiar landscapes of today. • 5,000 to 4,000 years ago – sea level was comparable with the present day. • 6,500 years ago – sea level was 8m above the present-day level. • 13,000 years ago – meltwater cut new valleys as ice melted from the area. Sea level was up to 45m higher than at present. • 27,000 years ago – the last advance of the Ice Age occurred as ice built in the Highlands. • 2 million years ago – the climate cooled and the Ice Age began.
Neogene	2–24	55° N	The climate cooled as the Ice Age approached.
Palaeogene	24–65	50° N	This was the age of explosive volcanic activity, when the North Atlantic Ocean was born and molten rock funnelled through the thinned and stretched Earth's crust to form a line of active volcanoes. The St Kilda, Skye, Rum, Ardnamurchan, Mull, Arran and Ailsa Craig volcanoes were all active during this time.
Cretaceous	65–142	40° N	Evidence from elsewhere in Scotland indicates that sea level was high at this time but no record is preserved in the Edinburgh area.
Jurassic	142–205	35° N	The climate was warm and humid.
Triassic	205–248	30° N	Desert conditions were widespread.

Period of geological time	Millions of years ago	Scotland's global position	Environments and events in and around Edinburgh
Permian	248–290	20° N	Volcanic activity early in this period gave rise to volcanic vents which contained fragments from deep in the Earth's crust and upper mantle.
Carboniferous	290–354	On the Equator	Throughout the Carboniferous – intrusions of dykes and sills including Hound Point, Salisbury Crags and Corstorphine Hill. • 315 million years ago – widespread forests gave rise to coal seams. • 320 million years ago – rivers deposited thick layers of sand. • 330 million years ago – forests were widespread across the area. • 335 million years ago – corals flourished in warm tropical seas. • 340 million years ago – oil shales were deposited in coastal lagoons. • 345 million years ago – small volcanoes erupted, recognised today as Arthur's Seat, Castle Hill and Craiglockhart Hill. • 354 million years ago – an arid coastal plain was periodically flooded by the sea.
Devonian	354–417	10° S	'Scotland' lay 10° south of the Equator as part of a great landmass. Desert conditions prevailed, although widespread river deposits were also preserved. Explosive volcanoes were erupting at this time.
Silurian	417–443	15° S	Final events unfolded in relation to the closing of the Iapetus Ocean as 'Scotland' and 'England' were finally united.
Ordovician	443–495	20° S	The Iapetus Ocean reached its widest point, and great thicknesses of sands and muds accumulated in the ocean deep. Rocks of this age make up part of the Southern Uplands to the south of Edinburgh.
Cambrian	495–545	30° S	No rocks of this age are recorded in the area.
Proterozoic	545–2,500	Close to South Pole	No rocks of this age are recorded in the area.
Archaean	Prior to 2,500	Unknown	No rocks of this age are recorded in the area. The Earth was formed 4,540,000,000 years ago.

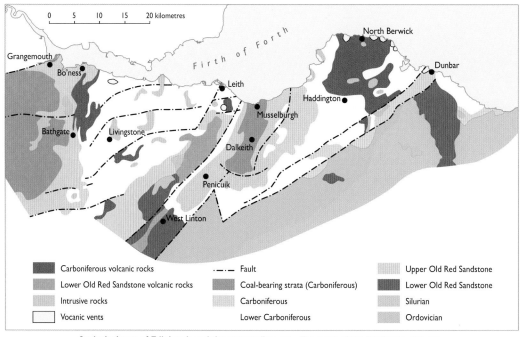

Geological map of Edinburgh and the surrounding area. The rocks of the Edinburgh area are defined by their age and type. They vary in age from the most ancient in the south of the area – the edge of the Southern Uplands – to rocks formed by events that happened at the end of the Ice Age. Rocks of Carboniferous age – sandstone, limestones, oil shales and coal-bearing strata – cover the majority of the Edinburgh area. Also from this time are significant volcanic outpourings of lavas and thick rock layers also of volcanic origin, known as sills. Salisbury Crags, the prominent rocky centrepiece of Holyrood Park, is an excellent example. Volcanic episodes were also a feature of the older Devonian Period. These strata were sliced by a series of faults or breaks in the rock, which provide evidence for significant periods of turbulence and instability to which this area was subjected in the geological past.

1
Edinburgh – birthplace of modern geology

An intellectual revolution took place in the second half of the eighteenth century in the university cities of Scotland. It was known as the Scottish Enlightenment. Gentlemen gathered weekly at dining clubs across Edinburgh to discuss the finer points of the natural sciences, economics, architecture and art. Out of this movement, the Royal Society of Edinburgh was founded in 1783 and received its Royal Charter the same year. We focus our attention on one particular group of luminaries: Adam Smith, the founding father of modern economics and author of the *Wealth of Nations*; Joseph Black, professor of chemistry at Glasgow and then at Edinburgh University; and Dr James Hutton, widely regarded as the father of modern geology. These men formed a close-knit group who established their own dining set – the Oyster Club. They met regularly, probably in what is now the Caves tavern in the Grassmarket. Their conversations were reportedly wide-ranging, 'but never didactic or disputatious'.

The Caves tavern in Edinburgh where the Oyster Club held their regular meetings.

John Kay was Edinburgh's 'society photographer' of his day. He captured the main characters of the era in a series of pen and ink drawings. James Hutton figured large among the luminaries sketched by Kay. In this illustration, the illustrious geologist is depicted beside a 'rock face' that figured a side view of some of his most celebrated contemporaries. Kay was a barber, miniature painter and a social commentator who observed the comings and goings of the elite of Edinburgh society. It was a great honour to be 'captured' for posterity by the redoubtable Mr Kay.

Dr James Hutton (left) and Dr Joseph Black in conversation – their sonsie faces 'full of science'.

There are many geological sites around Edinburgh that have a strong association with James Hutton. All these places helped to shape Hutton's evolving ideas on his theory of how the Earth works. Many of these localities are substantially unchanged since his day, so it's a fascinating experience to stand where Hutton would have stood over

two hundred years earlier and consider his ideas. So much has been revealed since then in terms of the basic tenets of geology that it is a stretch of the imagination to look at the natural world in the way that Hutton would have seen it.

Here are descriptions of two of the most famous sites associated with the life and work of James Hutton:

Siccar Point: this place is most famously associated with Hutton's developing ideas on the immensity of geological time. When Hutton, accompanied by John Playfair, visited Siccar Point on the Berwickshire coast in 1788, he saw that, in terms of geological time, they were 'staring into the abyss of time'. Red sandstones overlie upturned strata of what we now know to be rocks of Silurian age with a marked angular discordance, suggesting to Hutton that this complex relationship would have taken aeons of time to accomplish. Playfair later wrote 'what clearer evidence could we have had of the different formations of these rocks, and of the long interval that separated their formation'. We now know that separation between the deposition of the overlying sandstones and that of the Silurian beneath is around 50 million years. At the time of Hutton's visit, however, no-one could have come up with an estimate that would have even come close to this figure. The significance of this deduction resonates down the centuries and it forms one of the founding tenets of the science of geology.

Salisbury Crags: from his house on St John's Hill, Hutton would have overlooked what we now recognise as the wreck of an ancient volcano

Siccar Point. Layers of sandstones are draped over the upturned rocks of Silurian age. This discord is known as an 'unconformity'. To the untutored eye, it may look dull and unimportant, but to James Hutton, a man of undoubted genius, its significance was profound.

– Salisbury Crags and Arthur's Seat. In Hutton's day, all explanations of the natural world had to be grounded in the teachings of the Bible, with particular reference to the Flood mentioned in the Book of Genesis. But the idea Hutton advanced was something entirely new and ground-breaking. His theory was that the world functioned as a heat engine and some rocks had an igneous origin, where they found their place in the natural order of things in a molten and liquid form. This idea is universally accepted now, but it didn't command widespread agreement at the time. A battle between the Neptunists (those who believed in the Flood as described in the Bible) and Plutonists (who found evidence for the role of ancient volcanoes) raged for many decades.

The clues that back up Hutton's ideas are easily missed. At Hutton's Section (right) high up on Salisbury Crags, molten rock forced its way between existing layers of sedimentary strata, buckling the sandstone beds in the process. The sketch below by Hutton's companion John Clerk of Eldin shows the disruption of the sandstone beds. For some reason, the artist added figures at only a fraction of the correct size. This is not the only place where Hutton sought evidence for his ideas about the role of ancient volcanoes, but it is arguably the best known.

2
It's a dynamic Earth

Even in this short book, it is important to understand some of the bigger geological processes that shape the planet and the timescales over which they operate. Ours is a dynamic Earth, and change is a constant. We look at familiar landmarks like Arthur's Seat or Castle Rock and imagine they've been fixtures since time began. Formed 350 million years ago, they are ancient indeed, but the planet had a long history before these volcanoes were active.

So we need a better appreciation of geological time – the very concept that James Hutton pioneered. Most contemporary writers start the story of Scotland's history at best with the Ice Age or more usually with Scotland's first inhabitants. But the study of our geological heritage in Scotland tells a story that stretches back almost, but not quite, to the formation of Planet Earth.

The Earth's core is a raging 6,000°C and is the heat engine for our dynamic planet. Heat seeps from this central ball of iron and nickel to the enveloping mantle. This energy sets up a convection motion in the mantle that has a profound effect on the overlying layer – the Earth's crust. This outer layer of the planet is divided into seven large tectonic plates and many smaller ones that have moved across the surface of the Earth since earliest times. Over great expanses of time, the land that was to become Scotland moved across the globe, powered by a heat source deep within the planet.

The mantle 'flows' in response to the heat transferred from the Earth's core. This circulation powers the motion of the continents as they jostle and slide past each other. As continents collided, the layers of sand, mud, limestone and lavas that had built up on the sea floor were squeezed and folded into mountain ranges. The land that was to become Scotland has moved from a position close to the South Pole over a billion years ago to around 57° N of the Equator now. And that journey continues to this day as continents continue to move at a rate of around 3cm per year – the same rate at which your fingernails grow!

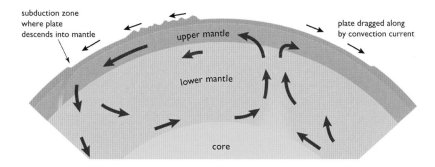

3
A disappeared ocean and baking deserts

The distribution of land and sea on Planet Earth was very different around 500 million years ago. Scotland was a small piece of a continent known as Laurentia, separated from England by the width of the Iapetus Ocean. But, over time, the ocean closed and these two component parts of what became the British Isles were united around 420 million years ago.

The youngest rock succession to have accumulated in the Iapetus Ocean is of Silurian age. A glimpse into this long-lost world is provided by rocks of the Pentland Hills, just to the south of Edinburgh. Any ocean, either ancient or modern, is characterised by deep waters, and the Iapetus Ocean was no different. Thick layers of mud were dumped in its murky depths. But as continents re-arranged themselves and the ocean closed, the nature of the sediments changed. We know that the different layers of sediment are interpreted as representing deep-water deposits, from the fossils found and their general character. But the most recent part of these rock layers is made up of strata that were laid down by a meandering river. So we are seeing the demise of an ocean that was at one time as wide as the North Atlantic.

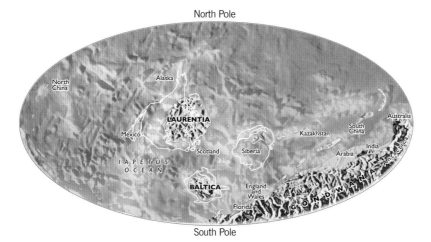

The geography of the world has changed throughout geological history, and it will continue to change into the future. This is how it looked 500 million years ago.

16

This once-mighty seaway had closed and, where deep seas formerly existed, the area was now dry land, albeit criss-crossed by flowing rivers. The rock faces that provide evidence for this interpretation are anything but spectacular, and can only really be appreciated by experts. But their significance is profound and signals the death of an ocean that existed for around 250 million years.

Devonian deserts

After the closure of the Iapetus Ocean, 'Scotland' lay close to the centre of a large continent that was positioned some 10° S of the Equator. North America and Greenland were Scotland's main bedfellows in this new continent.

A period of violent volcanic episodes followed: thick lava flows were erupted that built much of what became the Pentland Hills, the Braid Hills and Blackford Hill. Red sandstones of this age are present to the south of the Pentland Hills. These rocks are evidence of an unstable and immature land that experienced frequent earthquakes and volcanic episodes. The climate was hot, verging on semi-arid at times, with some seasonal rainfall. Sands were eroded from the surrounding mountains and blown by desert winds to lower ground that was scoured by flash floods. Complex fault systems then sliced the bedrock of this newly formed continent.

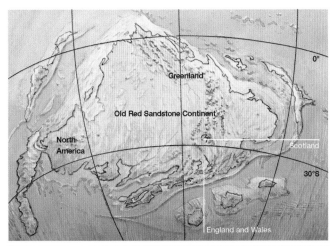

Below. The new landmass known as the Old Red Sandstone Continent was created by the closure of the Iapetus Ocean.

Right. This is an impression of what the landscape may have looked like during Devonian times around 370 million years ago. The mountains created by colliding continents were quickly eroded to rounder contours by the searing heat of the desert and by wind and rain. Rivers criss-crossed the lower ground, dumping sands and boulders transported from the mountainous areas upstream.

4
Roaring volcanoes and tropical lagoons

Rocks of Carboniferous age have a widespread occurrence across the Edinburgh area. 'Scotland' remained close to the centre of the ancient continent that comprised present-day North America, Greenland, Siberia and Scandinavia. This unexpected assortment drifted northwards, powered by the ever-restless mantle below. The land that was to become Edinburgh was now on the Equator. But it remained an unstable and restless place. The continent was subjected to earth movements that rifted and thinned the new continent. This led to a partial melting of the upper mantle leading to large quantities of molten rock being generated. Over the next few million years, this vast reservoir of molten rock was erupted at the surface or injected into the crust as a series of sills and dykes.

Many of the landscape features we recognise today in the Edinburgh area are the eroded stumps of the volcanoes that erupted during these turbulent times around 350 million years ago, and for

Pulses of molten rock that didn't quite make it to the surface solidified as sheets of rock sandwiched between pre-existing layers of sandstones or shales. These features are known as sills. Salisbury Crags is a particularly fine example of a sill.

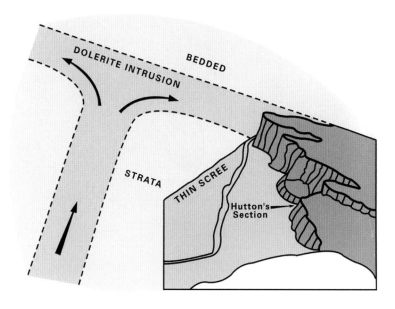

Salisbury Crags.

tens of millions of years thereafter. Arthur's Seat, Castle Rock, Craiglockhart Hill, Garlton Hills, and, further afield, North Berwick Law, are all prominent landmarks that were formed by eruptions during early Carboniferous times.

The East Lothian coast is another spectacular location to see the remnants of this violent and long-running volcanic episode. Along a 17-kilometre stretch between Dunbar and the off-shore island known

Above. The Edinburgh volcano erupted around 350 million years ago. This reconstruction captures the period when the volcano was active. It belched ash into the atmosphere and coughed up layer upon layer of lava from its main vent.

Right. Arthur's Seat with Lion's Head and Salisbury Crags in the foreground. The original structure of the Arthur's Seat volcano has been much modified by the passage of ice that has cut deep. This is fortuitous for geologists as it allows them to see the inner workings of a structure that would otherwise be hidden from view.

This unusual view from the south side of Edinburgh Castle illustrates the route the magma took. It cut through the sedimentary layers that were already in place (light-coloured strata that pokes through the grassy slope), pushing them to one side as the erupting molten rock swept towards the surface.

The stumps of ancient volcanoes are seen today as islands in the Firth of Forth. The Bass Rock is figured here.

as Fidra, an amazing variety of volcanic rocks and associated sedimentary layers are well displayed.

The ancient environment into which these volcanoes erupted was predominantly a series of shallow lagoons and lakes fed by meandering rivers. The rising magma encountered these waterlogged conditions during the final phase of its ascent from depth. The result of these frequent encounters was predictably explosive: eruptions were short-lived and extremely violent. Lake sediments built up around the volcanic vents, some yielding the fossilised remains of ferns and club mosses. These plant fossils are hugely significant in helping to reconstruct the ancient habitats that existed in the area some 340 million years ago.

An additional feature revealed along this coastline are lumps of the Earth's mantle and lower crust dislodged and transported from the depths by magma that streamed towards the surface. These give a unique insight into the composition of the lower reaches of Planet Earth at this time. Even the very deepest borehole doesn't come close to penetrating the crust and downwards into the mantle, so we're reliant on features like this to allow the deeply buried layers of the Earth to be directly sampled.

Above. North Berwick Law from the air.

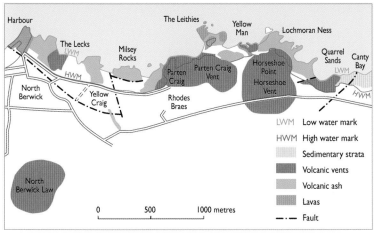

Left. This scenic coastline is studded with ancient volcanic vents, associated lava flows and accumulations of ash. This map illustrates the main elements of interest and the best way to access these locations. Much of the exposure is located between high and low water marks, so check tide tables before visiting.

Volcanic rocks are also present at other locations. On the south shore of the Firth of Forth, at a place called Hound Point, a thick sill of formerly molten rock dominates the sea cliffs. This coastal exposure is part of a bigger complex of igneous rocks known as the Midland Valley Sill complex that underlies West Lothian and connects with the Lomond Hills in Fife in a series of underground volcanic pulses (see overleaf).

Right. Hound Point. Molten rock was pushed into place between layers of pre-existing sedimentary rock. This is best seen just to the left of the hammer.

Opposite. Piles of burnt oil shale, known as shale bings, are characteristic of the West Lothian skyline.

Oil shales and James 'Paraffin' Young

The volcanoes around what would become Edinburgh eventually fell silent and a new order was established. A shallow lake, called Lake Cadell in recognition of the work of a famous Victorian Survey geologist, covered much of the area of the present-day Firth of Forth, City of Edinburgh and West Lothian. The level of the lake and its geographic extent rose and fell in synchronicity with the climate and there is evidence that it occasionally dried out completely. In the stagnant lower levels of the lake, thick layers of oil-bearing shales were deposited. The water here was putrid and home to algae and cyanobacteria, creating the ideal environment for the formation of oil-rich shale deposits. Shoals of fish thrived in the upper more oxygenated layers of the lake and these are well preserved as fossils in the rocks. Shrimps and other crustaceans were also abundant.

The commercial value of the oil shales of central Scotland was

exploited by chemist James Young in the late nineteenth century. For his efforts, he later became known as 'Paraffin' Young. In its heyday, this enterprise employed thousands of people in the local area. The shales yielded around 200 litres per tonne and, during the active life of the industry, it is estimated that around 75 million barrels of oil were produced. A range of products were produced as part of this industrial enterprise, including paraffin, lubricating oils and wax. The industry thrived for over 50 years and many millions of tonnes of shale were mined and treated. The waste products were piled high into hills of red blaes (or bings) that became a tangible reminder of the area's industrial past.

Below. James 'Paraffin' Young.

5
A cauldron of steam and lava

Thick bands of sandstone were also laid down at the same time as the oil shales. These layers of sand were dumped in a series of delta deposits by rivers carrying sediments from higher ground to the sea. The deposits were commercially worked for building stone some 340 million years later to construct Edinburgh's elegant New Town. The Craigleith, Hailes, Dunnet and Binny sandstones proved to be of particular commercial value.

Lizzie the Lizard and friends

Just outside the city limits lies one of the most remarkable fossil sites in Scotland. East Kirkton Quarry has been intensively studied over a long period and many remarkable finds have been made there. Interest was first triggered by a discovery made in 1985 by the professional fossil hunter Stan Wood. He found the remains of a four-limbed creature that represented one of the first animals to walk the Earth. The volcanoes of Edinburgh were still active in Lower Carboniferous times and the layers of rock preserved at East Kirkton are interpreted as an accumulation of strata within a series of lakes fed by a network of hot springs. The area was also ringed by active volcanoes. The energy that drove the hot springs system was provided by subterranean molten rock.

In this cauldron of steam and lava, many of Scotland's earliest weird and wonderful creatures lived side by side. The oldest known harvest spiders found anywhere in the geological record, and early forms of amphibians and reptiles, were all part of this diverse ecosystem. Other discoveries include giant scorpions, some up to 90cm in length; three different species of eurypterid, also known as sea scorpions; different types of fossil fish; and many plant remains. The lakes were fringed by thick forest dominated by club mosses and ferns. Evidence that these wooded areas were subject to frequent forest fires is provided by charred remains found among the plant fragments. The oxygen

content of the atmosphere was higher than it is today, making such occurrences more frequent.

The prize specimen to be discovered in the rocks of East Kirkton has become a national treasure. The fossil is probably better known by its nickname – Lizzie the Lizard. To the untutored eye, it looks like just a few smudges on a fist-sized piece of rough-hewn stone, but, to the expert, it is a treasure trove of scientific information. *Westlothiana lizziae*, to give this creature its scientific name, is thought to be an animal that was intermediate between amphibians and reptiles. Lizzie

Above. Fossil scorpion recovered from East Kirkton. Specimens of this type are rare indeed, and each new find helps to fill in gaps in our understanding of how these creatures evolved through time. This site has yielded more useful fossil scorpion material than any other site in Britain of this age. This specimen rejoices in the name of *Pulmonoscorpius kirktonensis*.

Left. Reconstruction of a scene in the East Kirkton area as it would have existed during early Carboniferous times. Lizzie the Lizard surveys her territory across the tropical rainforest towards the hot spring lagoon. In the distance are active volcanoes belching ash and steam.

lacks the diagnostic features of a reptile, but according to Professor Euan Clarkson, retired professor of palaeontology (fossil studies) at Edinburgh University, 'it appears to be a crucial "missing link" in the evolutionary chain from amphibians to the earliest true reptiles that appeared on the scene some 40 million years later'. Such was the importance of this find that it was bought for the nation after a public appeal. It is now on public display at the National Museum of Scotland in Chambers Street, Edinburgh.

The remains of the amphibian *Balanerpeton woodi* have also been found at the East Kirkton quarry. Around 30 complete or partial skeletons have been described from this location. This animal would have grown to about half a metre in length. The reconstruction shows what this animal would have looked like when it was alive.

6
Tropical seas and swamps

After a brief interlude a tropical sea environment developed, indicating another rise in sea level. Much of what is now the central part of Scotland was then under the waves. The reason for the large variations in sea level at this time, and also later in the Carboniferous Period, may have had its cause in the southern hemisphere. One idea suggests that the southern continents, as they existed then, had a variable cover of ice and snow. When the ice sheets expanded, they locked up the planet's freshwater resource, and sea levels fell across the world as a result. When the ice sheets melted, water was released into the oceans, and sea levels rose in response.

The warm shallow seas teamed with life, as tropical seas do today. A variety of species of coral, crinoids and shelled animals such as brachiopods inhabited this marine habitat. These strata are well seen along the East Lothian coastline around Barns Ness.

The rocks around Barns Ness lighthouse yield a fascinating variety of fossils, including crinoids (or 'sea lilies') and corals. Please limit any collecting to one specimen only.

Primitive coal swamps

By late Carboniferous times, much of the land that now lies between Edinburgh and Glasgow was covered by a tropical rainforest dominated by giant club mosses that grew to a height of 50m or more. The compressed remains of all the trunks, leaves, roots, seeds and branches of this 315-million-year-old biomass are recognised today as coal seams. These rocks were the foundation on which the Industrial Revolution was built some 250 years ago and significantly influenced where people lived their lives in Scotland and how they earned a living. So it is no surprise to learn that these strata are among the most intensively studied of any in the UK.

Let's have a closer look at these geological deposits that have had such a profound influence our way of life.

Reconstruction of a coal forest ecosystem.

Scotland lay on the Equator during late Carboniferous time, so the presence of an exotic habitat like a tropical rainforest comes with that territory. As during earlier times in the Carboniferous, sea levels fluctuated greatly, so coal swamps growing close to the coastal edge were periodically inundated by rising sea levels. Layers of sands and muds were dumped on the drowned forests and, over time, the organic remains were squashed under the burden of new sedimentary layers. As sea levels fell, forest areas would once again become established, only to be submerged under the waves as yet another rise in global sea level took place. This cycle was repeated on many occasions over long periods of geological time. So a sequence of coals, muds, sands and then a repeated sequence of the same layers built up throughout these times.

The regularity of the alternations of sand, mud and coal layers on the right-hand side of the photograph clearly shows how environmental conditions varied throughout this period. Coal represents periods of lower sea level as forests flourished and then the sea rose again to dump layers of sand and mud on top of the vegetation. Underlying the coals are fossil soils containing rootlets from the vegetation above. These regular cycles characterise all coal deposits in the UK and are known as 'cyclothems'.

More molten rocks

The final flourish at the end of the Carboniferous and into the Permian Period, around 285 million years ago, was the intrusion of more molten rock, this time in thick black layers. These dolerite intrusions cut through existing sediments at many sites across the city, such as Dalmahoy, Corstorphine and Ratho. These igneous intrusions stand proud in the landscape as they proved to be more resistant to the scouring effect of the ice than surrounding strata. Many have been, or are still being, worked to provide road stone. The close-grained dolerite is perfect for this purpose.

A dolerite quarry. The hard rock quarry has been exploited for road stone. Many are no longer worked.

7
Mind the gap!

One of the most intriguing aspects of the study of geology is the random nature of the evidence that is left for us to interpret. The geological record of the rocks has been referred to as 'pages from the Earth's autobiography'. To continue that analogy, many more pages were probably written, and some 'illustrated' by fossils, but were later torn up and discarded as the strata were eroded away and lost forever. The culprit was erosion – by ice, wind and water – that has cut deep, scoured and ravaged the landscape during various periods of our geological history. So in any given place in Scotland, there are huge gaps in the geological story. In Edinburgh, the Carboniferous Period, which opened around 354 million years ago and closed 64 million years later, is well represented. But after the intrusion of the dolerite sills late in the Carboniferous and into the early Permian Period (about 285 million years ago), there is little imprint of subsequent geological events on the land until the onset of the Ice Age that started around 2 million years ago.

Preserved elsewhere in Scotland's geological record is evidence for some momentous events: when the deserts sands of Permian times accumulated in great dunes, redolent of the present-day Sahara; or when the dinosaurs of Jurassic age stalked the land and patrolled the seas around the area we now recognise as Skye; or when the Atlantic Ocean opened and the volcanoes on Mull, Ardnamurchan and elsewhere roared in response to the shifting continents. None of these events are etched into the rocks of the Edinburgh area, or, at least, evidence of these occurrences has not survived.

So no one area of the country tells the full geological story, and each geographic region is important in piecing together the bigger narrative of the nation as a whole.

8
The Ice Age

Throughout this hiatus in the geological record, Scotland continued to drive northwards. In a matter of just over 300 million years, it moved from an equatorial climate to more northerly latitudes. This dramatic transition, where tropical conditions gave way to a full-on Ice Age, is etched into the landscapes of Edinburgh. Evidence of the

This is an iconic view of Scotland and one that many visitors to our shores will be familiar with. Its formation is rooted in two of the most primordial forces to shape our landscapes – fire and ice.

Castle Rock – moulded and shaped by the passage of ice. Edinburgh's other volcano, Arthur's Seat, is shown in the distance.

work of ice is clearly visible in many parts of the city and surrounding areas.

The Royal Mile, with Edinburgh Castle at one end and Holyrood at the other, is recognised as a classic crag and tail feature. Ice flowed from the west, but was forced over and round the hard volcanic plug upon which Edinburgh Castle sits. The softer (less resistant to erosion) Carboniferous-aged sediments that sat in the lee of the castle were thus protected from the full erosive force of the ice. As the ice moved over and round, it scoured and ice-plucked the volcanic plug, so that the western aspect of the plug was fashioned into the sheer rock face we clearly see from Princes Street today.

There are many other places where the erosive effects of the ice can be clearly seen, but none so historic as Agassiz Rock on the south side of Blackford Hill. In 1840 the celebrated Swiss geologist Louis Agassiz, who was in Scotland to attend a British Association meeting in Glasgow, chanced upon the site. Agassiz was well acquainted with the effects of glaciation in his native land. However, no-one had made the link between the erosive effects of ice and the dramatic landscapes of Scotland until Agassiz declared that the scratches on the rock he saw at Blackford Hill, and the landscape of Scotland generally, were

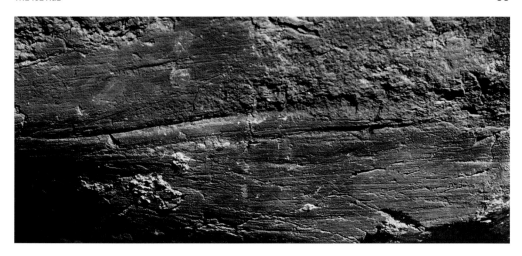

Scratches on the rock face at Agassiz Rock, Blackford Hill, Edinburgh.

'the work of ice'. The discovery made headline news in the *Scotsman* newspaper, which carried a full-page report describing Dr Agassiz's ideas.

Many landscapes in the Edinburgh area reflect the melting phase of the last glaciation, where water released from the ice coursed across the frozen landscape towards the sea. Deep grooves were carved in the land surface by this process. Carlops, just to the south of Edinburgh, is one of the best places to witness these effects. This valley is one of the most spectacular examples of the de-glaciation process – the end game of the Ice Age when the climate warmed and the ice melted away. In this instance, the ice cover was still present, but a raging torrent of meltwater flowed beneath the cover of the glacier. The channel at Carlops is one of many such features that criss-crossed the

Carlops valley was carved by water from the melting ice.

Raised beach at Cramond shore.

landscape, providing passage for the vast volumes of water released by the melting ice. These events took place some 14,000 years ago, close to the time when Scotland emerged from its glacial slumbers to become entirely free of ice. The final ice melt is dated at around 11,700 years ago.

The heavy burden of ice also had the effect of weighing down the land. So when the ice melted, the land was actually at a substantially lower level than it is today. The subsequent 'rebounding' of the land, now relieved of its ice burden, took many thousands of years to happen. It was like a cork bobbing back to the surface, but much more slowly. Millions of gallons of water had previously gushed into the oceans worldwide as the ice melted on a global scale, so seas rose to a higher level relative to the land. As a result, the coastal fringe was inundated. There is good evidence at Cramond, among other places, that sea level was a colossal 40m higher than it is today. At the same time, waves lapped against Calton Hill near today's London Road marking the edge of the Firth of Forth estuary as it existed 8,000 years ago. The presence of marine sands, silts and clays nearby allows this reconstruction to be reliably made. As the re-adjustment process took place, sea level fell relative to the land in fits and starts. The rebound stalled around 5,000 years ago and the sea cut a prominent notch into

Changing sea levels on the raised beach at Cramond.

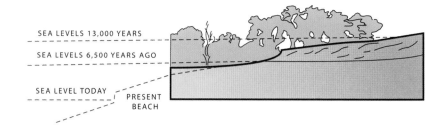

the land that fringed the coastline. It is marked today by the presence of a raised beach, again well illustrated at Cramond, close to, although 10m above, the high-water mark of the estuary today.

There were many hollows cut into the ice-scoured bedrock. Water soon filled those depressions to form a watery landscape of lochs and boggy areas. All except Duddingston Loch were subsequently reclaimed to create useful civic spaces. Perhaps the most well-known loch to be drained is the land now occupied by Princes Street Gardens. It was at one time a stinking sump for all the wastes produced by the good citizens of Edinburgh. Known as the Nor' Loch, it was filled in to create a recreational amenity area at the same time as the New Town was being constructed.

The Nor' Loch – in 1830 and today.

9
Building the New Town

The architectural brilliance of Edinburgh's New Town has been celebrated and widely acknowledged since it was completed in 1850. Edinburgh is often referred to as the 'Athens of the North'. The New Town was awarded UNESCO World Heritage status in 1985. What allowed this wonder to take shape was the ready availability of suitable

Craigleith Quarry (right) and Hailes Quarry were the largest man-made holes in the ground to be found anywhere in nineteenth-century Edinburgh. The technology to extract these massive blocks of rock was primitive, but ingenious. The quarrymen used their simple tools and hoists to manage truly monumental lumps of rocks that were then moved by horse and cart to construction sites around the city. The quarry was active from around 1750 until the early years of the twentieth century. The site has now been 'restored' and has found a second lease of life as the location for a supermarket.

building materials. Not all of the stone that built the New Town and other grand buildings of the city came from a location within the city boundary, but much of it did. Perhaps the finest of all was the Craigleith stone, used extensively in Edinburgh and exported to London and abroad in considerable quantities. It was used to construct many of Edinburgh's iconic buildings and structures, including parts of Edinburgh Castle, the Palace of Holyrood, Edinburgh University's Old College, the Royal Scottish Academy, the Old Royal High School, the National Gallery of Modern Art, Register House, the City Observatory on Carlton Hill and the Dean Bridge.

In terms of building stone imported to the city, brick-red sandstone from Locharbriggs near Dumfries was used to build the imposing Caledonian Hotel at the west end of Princes Street and stone from the Corncockle Quarry near Lochmaben was employed to construct the Royal Bank of Scotland building in Nicolson Street. And finally, the bright red sandstones of Triassic age from Longtown built the imposing and recently refurbished National Portrait Gallery in Queen Street.

Royal Scottish Academy, The Mound, Edinburgh.

Right. Old College, Edinburgh.

Below. Royal High School, Edinburgh.

10
Places to visit

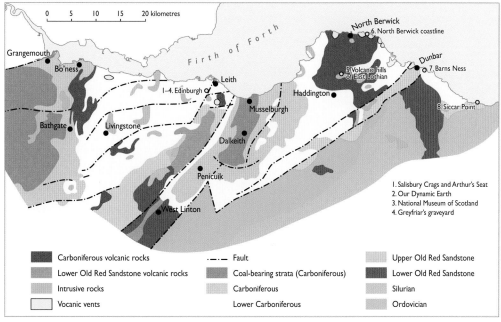

Map showing the locations of places to visit

Salisbury Crags and Castle Rock.

1. **Salisbury Crags and Castle Rock** are centrally located in Edinburgh. Features of interest here have already been described (see pages 18–20).

2. **Our Dynamic Earth**, located in the shadow of Salisbury Crags at the foot of the Royal Mile, provides an amazing family day out. It showcases some key global stories, among them the formation of Planet Earth, how plate tectonics work and management of the Earth's main ecosystems. The story starts with an exploration of geological time, assisted by holograms of the main players in this scientific investigation through the ages – Dr James Hutton, Sir Charles Lyell and Professor Arthur Holmes. The final messages are delivered in the dramatic setting of the Showdome, where an ever-changing series of short films addresses the scientific issues of the moment – from exploration of space to water management in the developing world.

Our Dynamic Earth.

3. **National Museum of Scotland** in Chambers Street displays an eclectic mix of artefacts and exhibits from Scotland and around the world. Geological stories and specimens are prominently displayed in two separate areas of the museum, including 'Beginnings', which tells the story of Scotland's geological evolution and also of James Hutton's contribution to the development of the science.

4. **Greyfriar's graveyard** at the foot of George IV Bridge in Edinburgh is where James Hutton, and other prominent figures from the Scottish Enlightenment, were laid to rest.

Above. North Berwick Law is the eroded remains of a once active volcano. It has been carved by ice into a classic 'crag and tail' form. A climb to the top of this local high spot is rewarded by panoramic views of the Lothians and the Kingdom of Fife. There is a convenient car park to the west of the Law.

Right. Traprain Law, near Haddington, was formed as a blister of lava that almost made it to the surface.

5. **Volcanic hills of East Lothian** – Garlton Hills, North Berwick Law and Traprain Law are located between Haddington and the coast at North Berwick and are well worth a visit. What remains are just the eroded stumps of former volcanoes that would have been active some 340 million years ago. OS Landranger sheets 66 and 67 will help you to navigate between these prominent landscape features.

PLACES TO VISIT

Above. Tantallon Castle sits on top of an ancient volcanic vent. The cliffs and foreshore are made from volcanic ash containing large blocks of rock that floundered within the volcanic vent.

Left. This natural wall of rock at North Berwick is a seven-metre-thick lava flow that sits on top of a layer of volcanic ash. The reddening of the ash layer is as a result of weathering in the tropical sun. These rocks were formed when 'Scotland' was situated at the Equator during Carboniferous times.

6. **North Berwick coastline** – pages 19, 22 and 23 give a description of the features of interest and a geological map of the site. The Land-ranger maps will help you to identify the best access points.

Barns Ness.

7. **Barns Ness** – good limestone exposures that yield a variety of fossils, including corals, brachiopods and the fossilised remains of a tree root, are to be found along the coastline near Barns Ness. Turn off the A1 just past the cement works and park near the beach. East Lothian Council has created a nature trail and published a leaflet for younger visitors, written by local children. Look out for Catcraig limestone kiln and the lighthouse as other points of interest.

Siccar Point.

8. **Siccar Point** – page 13 gives a description of the site. There is a car park within 200m of the coastline with information boards and a well-trodden path to the cliff top. Descending to shore level is hazardous as there is no safe path, but a good view of the key elements of the site is afforded from the cliff top.

Acknowledgements and picture credits

Thanks are due to Professor Stuart Monro OBE FRSE and Moira McKirdy MBE for their comment and suggestions on the various drafts of this book. I also thank Hugh Andrew, Andrew Simmons, Mairi Sutherland and Debs Warner from Birlinn for their support and direction. Mark Blackadder's book design is up to his usual high standard. Scottish Natural Heritage, in association with the British Geological Survey, published the 'A Landscape Fashioned by Geology' series that was the precursor to the new 'Landscapes in Stone' titles. I thank them for their permission to use some of the original artwork and photography in this book. David McAdam wrote the original text for *Edinburgh – A Landscape Fashioned by Geology* which influenced aspects of this book.

Picture Credits

Title page Christoph Lischetzki; 6 Permit Number CP16/085 British Geological Survey © NERC 2016. All rights reserved; 13 Alan McKirdy; 14 (upper) Alan Wilson/Alamy Stock Photo; 14 (lower) reproduced with the permission of Sir Robert Clerk Bt of Penicuik; 15 drawn by Jim Lewis; 16 drawn by Jim Lewis; 17 (upper) drawn by Jim Lewis; 17 (lower) Clare Hewitt; 18 Craig Ellery; 19 T.S. Bain/BGS; 20 (upper) Edinburgh 340mya © Gary Hincks; 20 (lower) cristapper; 21 Moira McKirdy; 22 Chris G. Walker; 23 (upper) ©Patricia & Angus Macdonald/Aerographica/SNH; 24 Permit Number CP16/085 British Geological Survey © NERC 2016. All rights reserved; 25 (upper) Lorne Gill/SNH; 25 (lower) © Dianne Sutherland; 27 (upper) NMS; 28 NMS; 30 SNH; 31 (upper) Permit Number CP16/085 British Geological Survey © NERC 2016. All rights reserved; 31 (lower) Permit Number CP16/085 British Geological Survey © NERC 2016. All rights reserved; 33 Krizek Vaclav; 34 Permit Number CP16/085 British Geological Survey © NERC 2016. All rights reserved; 35 (upper) John Gordon; 35 (lower) T.S. Bain/BGS; 36 (upper) T.S. Bain/BGS; 36 (lower) Craig Ellery; 37 (upper) © Edinburgh City Libraries; 37 (lower) Pablo Rogat; 38 Lorne Gill/SNH; 39 godrick; 40 (upper) Pete Spiro; 40 (lower) Heartland Arts; 42 christapper; 43 Tibor Bognar/Alamy Stock Photo; 44 (upper) Ulmus Media; 44 (lower) Lorne Gill/SNH; 45 (upper) Heartland Arts; 45 (lower) Alan McKirdy; 46 Alan McKirdy; 47 Alan McKirdy